WITHDRAWN

Science at Work: *Projects in Space Science*

ALSO BY THE AUTHOR

Science at Work: Easy Models You Can Make

Science at Work
Projects in Space Science

by Seymour Simon

illustrated by Lynn Sweat

FRANKLIN WATTS, INC.
845 Third Avenue, New York, N.Y. 10022

IOWA STATE TRAVELING LIBRARY
DES MOINES, IOWA

For Robert and his space-minded friends

SBN 531-01997-7
Copyright © 1971 by Seymour Simon
Illustrations copyright © 1971 by Franklin Watts, Inc.
Library of Congress Catalog Card Number: 70-171900
Printed in the United States of America

572632

MORNINGSIDE

Introduction

Space flight has become almost an everyday occurrence. Nowadays we rarely see a rocket launch on TV unless it's the beginning of a moon trip. Many other satellite and space-probe launches hardly rate a paragraph in a daily newspaper. Is it possible that space science is getting to be too commonplace for anything puzzling or surprising to happen?

Not likely. Not when close-up pictures of Mars reveal a surface unexpectedly like that of our moon. Or when what seems like a routine Apollo flight to the moon turns into a near tragedy with a mysterious explosion and a loss of oxygen supplies and power. Or when three Russian cosmonauts arrive back on earth dead after living safely in orbit for 24 days.

v

The world of outer space is still our most challenging frontier. This book will help you to broaden your knowledge of the principles and facts that space exploration is based upon. Using simple, easily available materials in and around your home, you can set up projects and investigate problems such as how a rocket works, how a meteoroid shield is designed, how a closed life-support system can exist in a spacecraft, how life can be detected on other planets, and how a radio beacon operates in space. As you are setting up these projects, keep in mind that the principles you see operating on a small scale also apply in complex space probes.

How can you see space science at work? By working at it yourself.

Contents

vii

part **1**

How Can We Get Into Space?

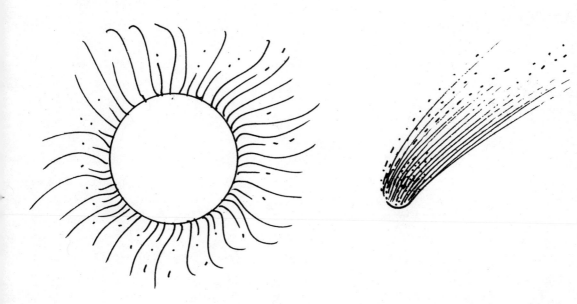

Experimenting with
a Projectile Launcher

YOU WILL NEED: A wooden board about 6 inches wide and 10 inches long, rubber bands, three ¼-inch wooden dowels about 8 inches long or three pencils, three small nails, and tape.

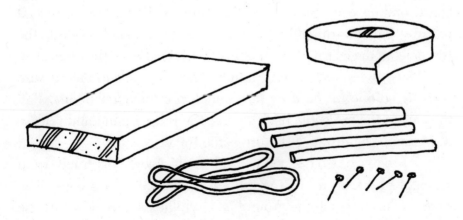

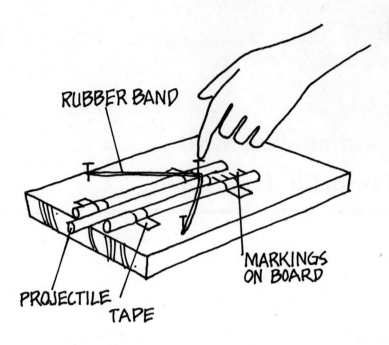

RUBBER BAND

MARKINGS ON BOARD

PROJECTILE

TAPE

WHAT TO DO: Tape two of the dowels lengthwise on the board. Leave enough room between them for the third dowel. Drive two of the nails about halfway into the board about 5 inches apart, one on each side of the two dowels. Drive the other nail halfway into the third dowel. Place a rubber band between the two nails on the board. Place the third dowel between the other two.

Place the launcher on a table 4 feet high. *Caution: Be sure there is no one in the area into which the launcher points.* Pull the nail on the projectile back against the rubber band and release it. Measure the distance the projectile travels.

WHAT TO LOOK FOR: When an object falls freely from a height of 4 feet, it hits the ground in about half a second. The time it takes to reach the ground is not influenced by how fast the object is traveling in a horizontal direction. You can show that this

4

is so by dropping an object from the table at the same time that you launch the projectile. Both will hit the ground at the same instant.

This means that if your projectile travels 7 feet horizontally in the time it takes to fall 4 feet, it travels the 7 feet in half a second.

You can use these numbers to determine the horizontal velocity of your projectile in miles per hour. Multiply the half-second and the horizontal distance by 7,200 (the number of half-seconds in one hour). This gives you feet per hour. Divide the number of feet by 5,280 to give you miles per hour. For example, a projectile that travels a bit more than 7 feet in half a second is traveling at about 10 miles per hour.

Change the amount you pull back on the rubber band. What effect does a longer pull have on the distance traveled? How does this affect the projectile's speed? Plot your finding along a graph such as the one shown below.

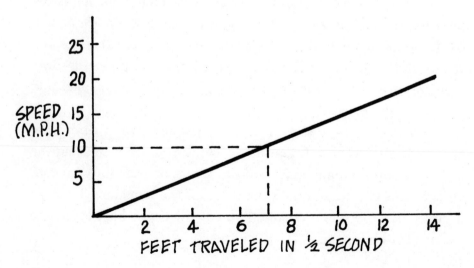

More Experiments with a Projectile Launcher

YOU WILL NEED: The same materials as in the previous experiment, more rubber bands, and dowels or pencils of different sizes.

WHAT TO DO: Set up the same projectile launcher as before, but try adding two or three rubber bands to the launcher. Next, use the same number of rubber bands but vary the size and weight of the projectile. In each case, launch the projectile from a height of 4 feet. Measure the horizontal distance traveled and convert to miles per hour.

WHAT TO LOOK FOR: When you use two or three rubber bands instead of a single rubber band, you are increasing the amount of force. Graph your results as shown on page 8. What conclusions can you reach about the effect of the launching force on a projectile's velocity? What do you predict will happen if you use more rubber bands? Try it and see.

Use the same number of rubber bands pulled back the same distance with different weight dowels. Measure the distance each dowel travels in half a second and convert to miles per hour as before. Which projectile travels the longest distance? Why do you think this is so?

When you apply the same amount of force to different objects, the objects speed up at different rates. The rate of increase of speed is determined by an object's mass. On the surface of the earth, an object's mass is the same as its weight. But mass remains constant even in the weightless conditions of an orbiting spacecraft.

Both of these experiments relate to Newton's Second Law of Motion. This law states in part that: 1) the greater the force on an object the greater its change in speed or direction; and 2) the greater the mass of an object the smaller the change in speed or direction. This relationship is less difficult to understand than it sounds. For example, it explains why it is easier to throw a light rock farther than a heavy one with the same amount of force. It also explains why you would use more force to throw the rock faster. Can you think of any other ways of illustrating the same idea?

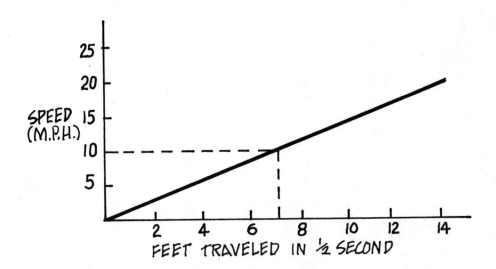

A Paper Match Rocket

YOU WILL NEED: Paper matches, aluminum foil, a balloon, a straight pin, and a paper clip.

WHAT TO DO: *Caution: Work this experiment in the kitchen sink because it involves lighted matches.* Cut out a piece of aluminum foil 1 inch wide and ½ inch longer than a match. Place a match on the aluminum foil so that its base is even with the bottom end of the foil. Place the pin against the match so that its point is at the head of the match. Wrap the match and the pin in the foil, folding down a double thickness around the head of the match. Carefully pull out the pin, leaving a clear channel under

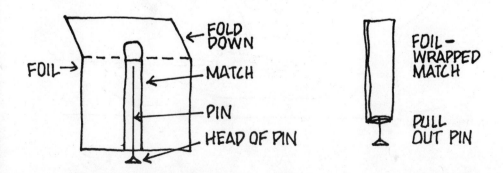

FOIL→
← FOLD DOWN
— MATCH
— PIN
— HEAD OF PIN

FOIL-WRAPPED MATCH

PULL OUT PIN

the foil from the head of the match to the base. Bend the paper clip into a handle to hold the match at an angle, with its head uppermost. Carefully hold a lighted match under the head of the foil-wrapped match until it ignites.

WHAT TO LOOK FOR: When the chemicals in the match head ignite, they change from a solid to a gas. The gas expands pushing against all sides of the foil. The expanding gases can only escape in one direction, through the channel. The match is forced in an opposite direction from the escaping gas.

This principle was first stated by Sir Isaac Newton about three hundred years ago. It is now called Newton's Third Law of Motion. He said that for every action there is an equal and opposite reaction. Think about what you saw the match do. What was the action? What was the reaction?

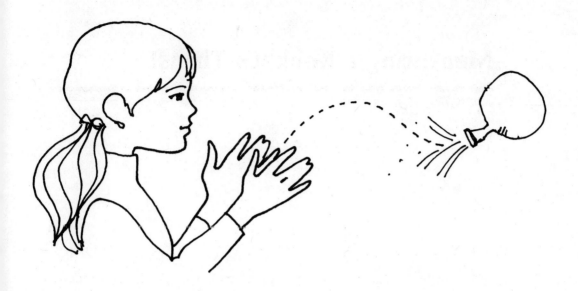

Blow up a balloon full of air, then let it go. In which direction does the balloon move? How can the Third Law of Motion help to explain what happens? How do you think you can make the balloon move faster and go farther? Try it.

You can make the rocket match go faster by using two wrapped matches instead of one. *Caution: Do not use more than two matches.*

Your match rocket used a solid chemical rocket fuel. Modern rockets use either solid or liquid fuels. Solid fuels are safer to store and handle, but liquid fuels offer more power from the same amount of weight. Liquid fuels can also be regulated in flight more easily than solid fuels. In most high-power space probes, liquid hydrogen and liquid oxygen, which help fuel to burn, are used to provide thrust. These liquids are made by cooling the gases to very low temperatures until they become liquids.

Measuring a Rocket's Thrust

YOU WILL NEED: A sensitive kitchen scale, an empty soda bottle with a cork to fit, a small pan, vinegar, water, bicarbonate of soda, measuring spoons, a paper napkin, a thumbtack, and string.

WHAT TO DO: Place the pan on the scale. Mix one tablespoon of vinegar with a cup of water and pour it into the soda bottle. Place the soda bottle on the pan and record the weight. Wrap a tablespoon of bicarbonate of soda in a square of a paper napkin. Insert the paper into the bottle and cork it immediately. *Caution: Do not push the cork in too firmly or the bottle may shatter.*

WHAT TO LOOK FOR: When the vinegar soaks through the paper and comes in contact with the bicarbonate of soda, a chemical reaction takes place. A gas, carbon dioxide, is produced. The expanding gas pushes the cork out of the bottle with a pop.

Watch the scale as the cork pops out. It will show the amount of thrust provided by the cork. Why does the cork's thrust push the bottle downward? This experiment measures thrust in much the same way that the thrust of large rockets is measured in static-

12

test stands. What are the advantages of a static test over launching a rocket each time you want to test its thrust?

Try experimenting with different amounts and proportions of the chemicals. Can you increase the thrust of the rocket? Push a thumbtack into the top of the cork to increase its weight. Does this affect the amount of thrust? In what way?

If you do not mind cleaning up a little mess on the bathroom floor, try this. Tie a string around the neck of the bottle and another string around the body of the bottle. Pour in the vinegar. Tie the two strings on the shower head over the bathtub so that the bottle hangs sideways and swings freely. Insert the paper with the bicarbonate of soda and cork the bottle.

What do you expect to happen when the cork pops out? Try it and see.

Computing Altitude and Mass Ratio of a Water Rocket

YOU WILL NEED: A water rocket and hand pump (sold in many toy or hobby stores for under $1.00), water, a scale, and a sweep-second-hand wristwatch.

WHAT TO DO: You prepare a water rocket for launch by first filling it about halfway with water and then using the hand pump to load the rocket with compressed air. Do not use more pressure than is written in the directions accompanying the rocket. Too much pressure may cause the plastic rocket to crack. Launch your rocket in an open field so that you can easily retrieve it and so that it will not break on impact. *Caution: Never point the pumped-up rocket at anybody or at any part of your own body.*

WHAT TO LOOK FOR: One method of finding out the altitude your rocket reaches is to time the interval in seconds between launch and impact. Then use the formula: $H = 16(\frac{t}{2})^2$ where H is the altitude, and t is the time in seconds. For example,

14

let's say it took 4 seconds from launch to impact. You would first divide 4 by 2 which would give you 2. You would then square 2 to give you 4 and multiply this by 16. In this example, your rocket would have reached an approximate height of 64 feet.

The mass ratio of a rocket is one way to measure the final speed it will attain. The higher a rocket's mass ratio, the faster its speed. To find the mass ratio of your water rocket, weigh it when it is filled with water before a launch and then weigh it after the water is expelled. The mass ratio is equal to the weight of a rocket loaded with fuel divided by the weight after the fuel is used. For example, suppose your rocket weighs 1 pound when loaded and ½ pound afterward. You would divide 1 by ½ to give you a mass ratio of 2. Would lighter weight or heavier weight fuels let you attain faster speeds? Is water a good propellant?

Try using different amounts of water, but be sure not to exceed the limit on pressure. Would twice as much water launch a rocket to a higher or a lower altitude? What influence does the air pressure have? Use fewer strokes of the hand pump and compare. What combination of water and pressure gives you the longest flights?

Water rockets are also made in two-stage models. If you can purchase one of these, you have even more of a chance to experiment. Does the same amount of water split between two stages instead of in one stage result in longer flights? What special problems do two-stage rockets present? What are the advantages and disadvantages of each kind of rocket?

Testing for Pure Fuels
in Spacecraft

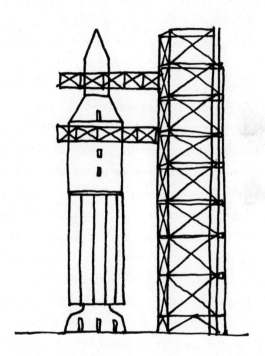

YOU WILL NEED: Copper sulfate crystals (found in many chemistry sets and can also be purchased in a drugstore), a pyrex dish for use in an oven, a small jar with a cover, and half a small jarful of alcohol. *Caution: Alcohol is inflammable and should not be allowed near an open flame. Keep the alcohol covered.*

WHAT TO DO: Place a small amount of the copper sulfate crystals in the pyrex dish. Place the dish in an oven and set the temperature at about 500° F. Check every five or ten minutes until the chemical turns a pale gray color. Remove from the oven, and turn the oven off. Place the copper sulfate in a small jar and keep it tightly covered. Now use the copper sulfate to test for the presence of water in the alcohol. Place some of the copper sulfate in the alcohol sample. Cover the jar and shake.

WHAT TO LOOK FOR: If the alcohol contains water, the grayish-colored copper sulfate will turn blue. If no water is present, the copper sulfate will remain gray.

Rocket fuels must be made of very pure materials. Even small amounts of impurities, such as water, may cause problems in fuel burning. A reduced amount of rocket thrust may result from these impurities. This variation can easily throw a rocket off course and prevent it from accomplishing its mission.

Almost all alcohol has some water mixed in with it. In fact, alcohol is sometimes added to the gasoline in an automobile. The alcohol, sold under the name "dry gas," mixes with any water in the gasoline and prevents ice from forming in the fuel lines.

Of course rocket fuels may contain impurities other than water. Can you think of how you could get rid of solid impurities such as dust particles? How are these particles removed from the gasoline in an automobile? Try constructing a similar filter for a dirty alcohol sample.

18

Escape Velocity

YOU WILL NEED: Poster cardboard or heavy construction paper, tape, clamps, two piles of books at least 1 foot high, and marbles.

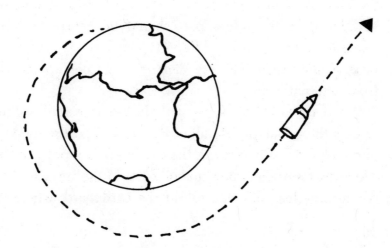

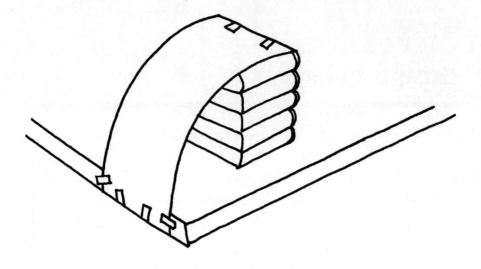

WHAT TO DO: Fasten one end of the poster cardboard to an edge of a table with tape. Bend the cardboard in a smooth curve and fasten the other end to the tops of the book piles with the clamps. The lower part of the cardboard should be nearly vertical while the upper part should be nearly horizontal. Try releasing marbles from different points on the cardboard and observe how they fall. Try flicking a marble up the cardboard from different points.

WHAT TO LOOK FOR: The slope of the cardboard represents the gravitational pull of the earth. Because the lowest part of the cardboard is vertical, the marble falls rapidly at that point. This area represents the situation close to the surface of the earth. A marble released at the top of the cardboard, where the curve is less, will roll slowly downward and pick up speed gradually. This upper area represents a distant point in space where the earth's gravitational pull is very slight.

What is the difference in the marble's acceleration at different points on the cardboard? How do the different rates of acceleration relate to the different amounts of gravitational pull? Do you think a heavy marble will accelerate at a different rate than a light marble? Is the same thing true of all falling bodies?

Flicking the marble up the slope from the lower edge simulates the launch of a spacecraft. Try flicking the marble at different speeds. What effect does that have on the marble's path? Where is the most energy needed to pull away from the gravitational pull of the earth — near the surface or far away?

Starting a marble up the slope from a spot near the top of the cardboard represents the situation of launching a craft from an orbiting space station. What are the advantages of a launch point such as that?

How could you set up another cardboard to simulate how the moon's gravitational field takes over from the earth's gravitational field? How does the second model help to explain why a spacecraft's main-thrust rocket engines can be shut off during much of the trip from the earth to the moon? At what point can you say that the ship is coasting downward to the moon?

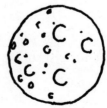

Stabilizing a Spacecraft

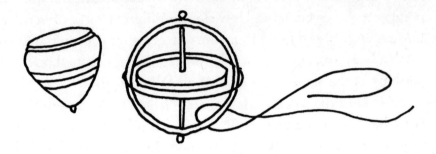

YOU WILL NEED: A toy top, a toy gyroscope, and string.

WHAT TO DO: Stand the top on end and let go. Now set it spinning and set it on end. Observe the difference. Start a toy gyroscope spinning. Try balancing it on a taut string or on the edge of a table. Hold one end of the spinning gyroscope and try to turn it in another direction.

WHAT TO LOOK FOR: Spacecrafts can tip in three different ways when traveling forward. You can feel this kind of tipping when you try balancing a broom upright in the palm of your hand. In a spacecraft the rotating motion is called *roll*. A side-to-side motion is called *yaw*. An up-and-down motion is called *pitch*. Besides being upsetting to any astronauts aboard a spacecraft, these motions make flight guidance difficult and affect the way certain instruments work.

The spinning top or gyroscope illustrates one way to keep a spacecraft in a stabilized flight path. A spinning object tends to remain pointed in the direction of its axis. Because of this principle, spinning along an axis counterbalances any forces that might cause the spacecraft to wander off course. For the same reason, a rifle barrel is grooved so as to spin the bullet as it emerges. Can you think of the reason the earth does not wobble very much in its orbit around the sun?

What happens when you try to move a spinning gyroscope from its axis of rotation? Because a spacecraft needs to be stabilized in each of three directions, how many gyroscopes would you need aboard to signal the controlling mechanisms of the craft? How would you set them up in relation to each other?

If you have a bicycle wheel to use you can feel a form of stabilization on your own body. Set a piece of broomstick on each end of the wheel axle to use as a handle. Hold on to one handle and spin the bicycle wheel as fast as you can. Now try to move the wheel in different directions. What happens? Try sitting on a piano stool or a swivel chair with your feet off the ground. Twist the spinning bicycle wheel. What happens? Look at the heavy wheel of a toy gyroscope. Can you think of how you could make a bicycle-wheel gyroscope more effective?

Heat Problems During the Return to Earth

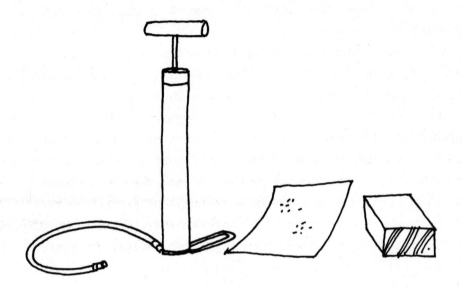

YOU WILL NEED: A block of wood, sandpaper, an air pump.

WHAT TO DO: Rub the sandpaper over the block of wood. Touch the block of wood with your fingers. How does it feel? Rub the sandpaper more quickly over the wood. Do you notice any difference in how it feels? Pump the air pump rapidly for a minute or two. Touch the bottom of the pump where the air is compressed. How does it feel?

W H A T T O L O O K F O R : When two surfaces rub together, they become heated because of the friction between them. The faster and the harder the surfaces rub, the more heat is generated. This effect takes place even when an object moves through the air. Can you think of why we normally do not feel the heat of air friction?

Meteoroids are chunks of iron or rock that travel through space between the planets. When one of these chunks enters the earth's atmosphere, it begins to heat through friction. It begins to glow red hot and melt. The glowing gases that can be seen form a meteor. Some people call meteoroids shooting stars when they see them move across the night sky. Most meteoroids burn up and turn into gas before they get within 50 miles of the earth's surface.

A space capsule also travels very quickly. What do you think would happen if it were to plunge through the earth's atmosphere at top speed? Even airplanes, which travel at much slower speeds than spacecraft, experience temperature increase at their surfaces. Moving at 700 miles per hour, about the speed of sound, the temperature increases by about 95° F.

An orbiting satellite travels at about 17,000 miles per hour. If the satellite returned to the earth's atmosphere at that speed, surface temperatures might go up to about 3,000° F. A spacecraft coming back to the earth from the moon may be moving at speeds in excess of 25,000 miles per hour. Surface temperatures at that speed in the atmosphere may go above 10,000° F.

Make a list of different materials used at high temperatures around your home such as in cooking utensils. Try to find out their melting temperatures from an encyclopedia or some other source. Which materials are likely to be used in the nose of a spacecraft? How do the passengers in such a craft live through the heat? Look at the next project to see how this problem is solved.

Designing a Heat Shield

YOU WILL NEED: Two pyrex test tubes or two baby (nursing) bottles, water, two thermometers, candle wax, an electric hot plate, a pan, and a double boiler.

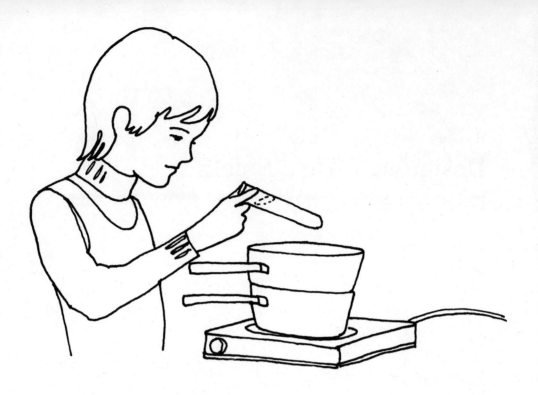

WHAT TO DO: Melt some candle wax in the top of a double boiler. Dip one of the test tubes or bottles in the melted wax and remove it. Allow the wax to harden. Do this several times until a thick wax coating is built up. Fill each of the test tubes with an equal amount of cold water. Place a thermometer in each. Stand the test tubes in a pan of water on a hot plate. *Caution: Use the hot plate with care. Be sure to turn it off when you are finished with it.* Heat the water in the pan. Note the temperature in each test tube every thirty seconds.

WHAT TO LOOK FOR: Why does the water in the wax-coated test tube remain cooler than the water in the uncoated test tube? What happens to the wax as the water heats up? Could this kind of a heat shield be used more than once?

28

In the same way that the water remained cooler while the wax melted away, a spacecraft will remain cooler if it has a shield that melts away because of the atmosphere's friction. The material that melts away is called an ablation material, and the shield an ablation shield. The word "ablation" means removal. Look at newspaper or magazine photos of a spacecraft and try to find the ablation shield. What do you think would happen if the shield did not enter the atmosphere first?

In the capsules used in Mercury, Gemini, and Apollo, the ablation material coats the flat, blunt end. Apollo uses a heat shield made of a special plastic held together by steel strips.

Some of the heat-resistant materials discovered as a result of space research have been used in other ways. For example, pyroceram is a material used to make cookware that not only can stand high temperatures, but can also be put into water while red hot and still not crack. This material is sold under the commercial name of Corningware. Even specially treated glass such as pyrex could not withstand such sudden heating and cooling. The nose cone materials of spacecraft are subjected to still more severe conditions.

29

IOWA STATE TRAVELING LIBRARY
DES MOINES, IOWA

Experimenting with g-Forces

YOU WILL NEED: A heavy rubber band, a small lead sinker or other weight, a wide-mouthed jar, and a pencil.

WHAT TO DO: Cut the rubber band in half. Tie one end to the lead weight. Place the lead weight in the jar and lay the pencil across the top of the jar. Tie the other end of the rubber band around the pencil so that the weight is suspended about midway in the jar. Grasp the jar with both hands and move it sharply upward. Wait until the weight is again midway in the jar and this time move the jar sharply downward.

WHAT TO LOOK FOR: When the weight is hanging midway in the jar, the pull of gravity downward equals the tension on the rubber band. What happens to the length of the rubber band when the jar is moved upward? What happens when the jar is moved downward?

The lead sinker can represent the weight of an astronaut. The jar moving suddenly upward represents the instant of a launch. The apparent weight of an astronaut increases during the launch. We say that g-forces are acting upon him. A force of 2-g would increase his weight two times.

31

In an actual spacecraft launch, the g-forces are small at first. But they rapidly build up to about 7-g within two minutes. If an astronaut weighs 150 pounds on earth his weight at this time would be over 1,000 pounds.

At the end of a space flight, these forces may reach an even higher level than at launch, perhaps as much as 20-g. Experiments have shown that a human body can stand as much as 45-g — but for only a split second.

How is the cabin of a spacecraft designed to protect an astronaut from the effects of g? Consult the works listed in *Books for Reading and Research,* at the end of this book, to find out some of the ways.

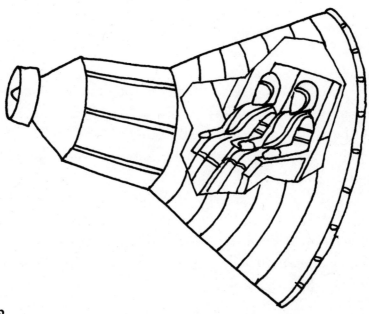

part 2

How Can We Exist
In Space?

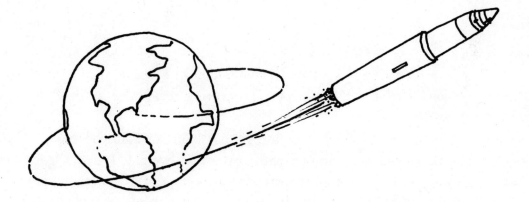

Weightlessness Aboard
A Spacecraft

YOU WILL NEED: A wide-mouthed jar, a heavy lead sinker, a rubber band, a pencil, a spring scale, and a friend to help you.

WHAT TO DO: Cut the rubber band so that you have a single strand, and tie one end to the lead weight. Put the lead weight in a jar so that it rests on the bottom. Tie the other end of the rubber band around a pencil that has been placed across the top of the jar. The length of the rubber band should be adjusted so that it supports most of the weight of the sinker, which just barely touches the bottom of the jar. Hold the jar as high as

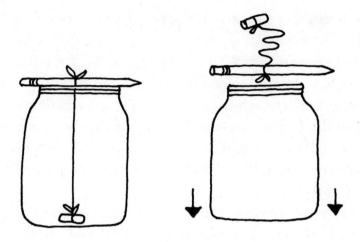

possible above a soft surface such as a mattress. Drop the jar and have a friend try to catch it before it hits.

WHAT TO LOOK FOR: Weightlessness does not mean that there is no gravity. In space, weightlessness occurs when a gravitational force is counterbalanced by the rocket's motion. This happens during the time when the engines have stopped and the satellite is coasting around the earth. What happens to a weight in a rocket when the engines are firing?

The lead sinker becomes weightless during the time it is falling. Since the sinker is weightless, the stretched rubber band will pull it out of the jar while it is falling.

You don't think that the sinker is weightless while it falls? Try this. Obtain a spring scale and hook the sinker onto it. Holding on to the scale, drop your hand rapidly. What happens to the weight shown on the scale during the time your hand drops? If your hand could drop fast enough, the weight would go to zero.

You can feel the same sensation of losing weight in a high-speed elevator when it starts to drop, or on a roller coaster during the big downward curve. Your body feels as if it is floating upward and you have to grab something to hold you down.

What happens when an astronaut in a weightless condition wants to drink from a glass? How could this be accomplished? What would happen if the astronaut pushed against one of the walls of the spacecraft? What other kinds of things happen during weightlessness in the spacecraft?

If air were not circulated by fans inside a spacecraft, an astronaut would die of suffocation in his own exhaled breath. Air currents on the earth come about as a result of heated air becoming lighter and moving upward. But if the air in a spacecraft is weightless, then the carbon dioxide an astronaut breathes out would stay around his face and choke him to death. What would happen if you lit a candle in a weightless state? You can see that weightlessness makes a pretty good fire extinguisher.

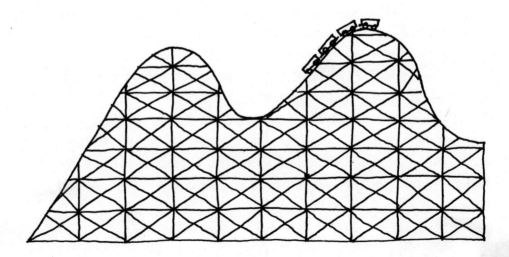

Making Artificial Gravity

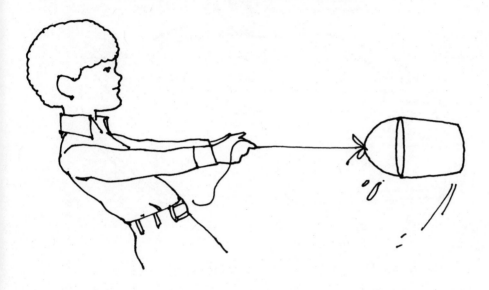

Y O U   W I L L   N E E D : A bucket, string, water, a phonograph turntable, a cardboard strip 1 inch wide and long enough to go all the way around the turntable, tape, and a marble.

W H A T   T O   D O : *It is a good idea to do this experiment out-of-doors — just in case!* Make sure that the string is strong enough to withstand a sharp pull. Tie the string securely around the handle of the bucket. Place some water in the bucket. Holding the string firmly, start to swing the bucket around in a circle. If you swing the bucket fast enough, the water will be forced against the bottom even when the bucket is upside down.

37

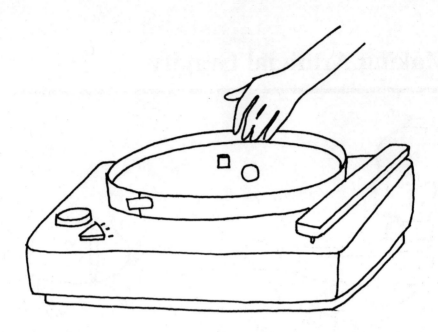

To see another example of this kind of outward force, attach the cardboard strip with tape around the edge of the phonograph turntable to form a rim. Start the turntable rotating and place a marble near the center pole. Note which way and how fast the marble rolls. Try spinning the turntable at different speeds to see the effect.

WHAT TO LOOK FOR: Of course, you really did not make gravity in this demonstration. You substituted another force in its place. This outward pushing force is called centrifugal force. It is caused by the rotation or spin of an object. What happens to the amount of this force when you make the turntable move faster? In relation to the earth's gravity, what part of the turntable represents up? What part down?

Scientists are thinking of constructing a space station in the form of a kind of spinning doughnut. How would you construct rooms in this kind of station? Where would the floors and ceilings have to be located?

Are there any other ways man might move around a space station without floating? What would be the advantages and disadvantages of using magnetic shoes to keep an astronaut on the floor?

At takeoff, an astronaut must withstand many g-forces. In space, he must get used to zero-gravity or weightlessness. He also has to withstand vibrations, accustom himself to special foods and to being confined for long periods in a small space. How would you react if you were closed up in a small dark space for a long period of time? How would you behave when under constant observation by monitoring stations on the earth? What kind of personality qualities should an astronaut have? Do you think you would make a good astronaut?

Measuring Speed Changes Aboard a Spacecraft

YOU WILL NEED: A lead weight, string, a clipboard, paper, a ruler, a pencil, and a friend to help with this project.

WHAT TO DO: Insert a piece of paper onto the clipboard. Tie one end of the string to the lead weight. Place the other end of the string under the clip so that the weight hangs just below the bottom of the board and is able to swing freely. Hold the board in an upright position, allowing the weight to stop swinging and come to a rest. Suddenly start walking rapidly, carrying the board. Observe what happens to the weight and the position of the string.

WHAT TO LOOK FOR: How would you know when a craft changes its speed in space? On the earth we can use an instrument that moves when in contact with the ground, water, or air like a speedometer. But space has none of these elements that the craft moves past.

Newton's First Law of Motion states that an object remains at rest, or in the same kind of motion in a straight line, unless acted

40

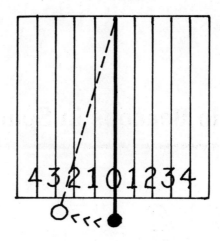

upon by some other force. The force that acts to keep an object at rest or in motion is called inertia. An increase in the speed of motion is called acceleration. The instrument used in space to measure a change in speed or direction is called an accelerometer. It uses the principle of inertia.

Now try to use your clipboard to demonstrate how an accelerometer functions. Wait until the weight stops swinging. Move slowly. Is the weight deflected to the same point as when you moved rapidly? Rule vertical lines on the paper 1 inch apart. Starting from the middle, number each line consecutively to the edges of the paper. You can see that the farther away from the center the weight moves, the greater the change in speed. How can you tell about changes in direction with the accelerometer?

Have a friend on a bicycle or on skates hold the accelerometer as he moves at a constant speed. How does the weight hang now? How does an accelerometer's reading differ from a speedometer's reading?

41

Radio Beacons in Space

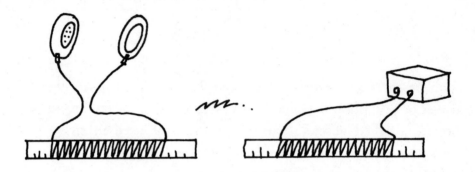

YOU WILL NEED: A toy train transformer (see below in *What To Do*), bell wire, two wooden or plastic rulers, and earphones.

WHAT TO DO: Wrap about 100 turns of the bell wire around each of the rulers. Leave about 24 inches of wire loose at each end. Connect the leads (wires) of one ruler to the terminals of the toy transformer. Supply the coil with 2–6 volts A.C. (Note: Make sure that you are not using a transformer-rectifier that supplies D.C. rather than A.C.) Connect the second set of leads from the other ruler to a set of earphones. Hold the coils a foot apart. Turn the second coil until the hum in the earphones is at its weakest. Move the first coil in a different direction. Line up the second coil for the weakest hum again.

WHAT TO LOOK FOR: The first coil is acting as a radio-wave transmitter. The radio waves are broadcast at the frequency that electricity comes into your house power supply, usually 60 cycles per second. The second coil acts as the antenna for a radio-wave receiver. It picks up the 60-cycle hum faintest when it is in what position relative to the transmitter? The position where reception is faintest is called the null situation.

Can you think of how picking up the direction of a transmitter whose location is known can help you find out your own location? Special navigation transmitters are called radio beacons. How many such "fixes" of a known transmitter would you need to locate yourself exactly? How could you "home" in to a radio beacon? What would tell you when your craft leaves its course?

If you have a portable radio, you can experiment with the same principle of radio location. Tune in a distant station with a weak sound. Rotate your portable in a circle. The sound should be faintest at one point. This is the null position. Modern portable radios usually use a ferrite-stick antenna. In that case a null is obtained when the station is lined up with the stick. Usually, the stick is in line with the length of the portable radio. Special radios with rotating antennas are often sold to small boat owners for this kind of radio navigation aid.

Meteoroid Shield
for a Spacecraft

YOU WILL NEED: Pebbles and rocks of various sizes up to about 1½ inches in diameter, plaster of Paris, a flat box or basin, a sheet of aluminum foil, and a sheet of plastic food-wrap.

WHAT TO DO: Cover the bottom of the flat box with a layer of wet plaster of Paris to a depth of at least 2 inches. Drop different-size rocks from a height of 1 foot onto the surface of the plaster. Increase the height to 2 feet, and drop other rocks. Also try throwing the rocks at various angles. Finally, use a sheet of aluminum foil as a shield to cover the plaster. Keep the foil away from the surface of the plaster by bending it over the sides of the box. Try dropping rocks onto the foil shield.

WHAT TO LOOK FOR: Meteoroids are chunks of rock or metal speeding through space. The impact of the rocks on the plaster of Paris is similar to the impact of meteoroids on the surface of the moon. Observe the kinds of craters made by dropping the rocks. Do they look like the craters you have seen in photographs of the moon's surface?

44

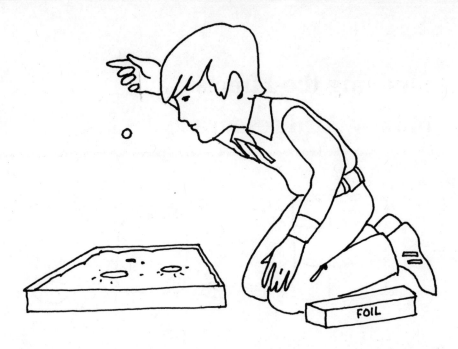

Is a crater the same size as the rock that made it? Does the size of the crater depend upon the velocity of the rock when it hits the surface? How does the angle of impact affect the appearance of the crater?

Spacecraft in flight need some protection from meteoroid strikes. A thin layer of metal or other material held slightly away from the hull is one method of slowing down the meteoroids and lessening their effect. Did the foil decrease the size of the craters? What does this show about the effectiveness of such a shield? Can you think of any drawbacks to the use of this kind of shield?

Try using a sheet of plastic food-wrap as a shield instead of the foil. Is it more or less effective? A good shield will absorb most of the meteoroid's energy of motion before it reaches the hull. Can you think of any other materials that might be used? Remember that a light weight is an important requirement for spacecraft use.

Studying the Effects
of Low Air Pressure

YOU WILL NEED: A 1-gallon can that has a tight-fitting screw cap, water, an electric hot plate, a balloon, and a wide-mouthed pyrex jar with a top. *(The jar must be made of pyrex because it will be heated.)*

WHAT TO DO: Make sure that the can is rinsed and clean. Place about one-half cup of water in the can. Heat it over the hot plate until the water is boiling rapidly for a few minutes. *Caution: Use the hot plate with care. Turn it off as soon as you have finished with it.* Remove the can from the heat and cap it promptly. Allow the can to cool. What happens to the can?

For the second part of this demonstration, blow up the balloon just enough to separate the sides, and close off the balloon with a knot. Place a small amount of water in the pyrex jar and heat it over the hot plate. When the water is boiling rapidly, drop the balloon into the jar and cover the jar tightly. Remove the jar from the heat and let it cool off. What happens to the balloon inside the jar?

WHAT TO LOOK FOR: At sea level on the earth, air exerts a pressure of 14.7 pounds against every square inch of our bodies. This means that many tons of air are pressing against our entire bodies' surfaces all the time. But an equal pressure of air and fluids inside our bodies pushing outward prevents us from being crushed under this great weight.

The demonstrations in this project show what would happen if this balance of pressure is upset. The steam from the heated water drove most of the air from the can. When the can was capped, the water inside condensed, leaving a much lower pressure. How does this help to explain why the walls of the can caved in?

Much the same thing happened inside the pyrex jar. What happened to the pressure inside the jar when the steam cooled? Why did the balloon in the jar expand at that point?

What would happen to man on the moon or in space if he had no protection from the lack of air pressure around him? What are some ways that man can be protected in low-pressure surroundings? Deep-sea diving suits are similar to space suits in some ways but different in other ways. Can you compare them?

47

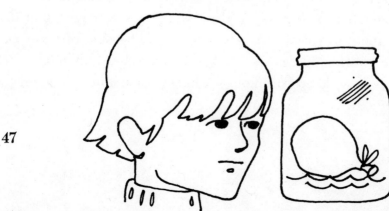

Water and Water Vapor
in a Spacecraft

YOU WILL NEED: Ice cubes, water, colored ink, a wide-mouthed jar and cover, and a saucer.

WHAT TO DO: Fill the jar with ice cubes. Add water to fill in the spaces, then add a few drops of ink and cover the jar tightly. Place the jar on a saucer. Let the jar and saucer sit on a table for a few minutes. Observe what forms on the outside of the jar.

WHAT TO LOOK FOR: The air in your room has water vapor in it. But the water vapor is usually not present in amounts large enough to bother you. If there were no way to get rid of the water vapor in a spacecraft, it would rapidly become very uncomfortable and finally dangerous. You can see how this might feel by sitting in a closed bathroom while hot water runs continuously from the shower.

48

The water droplets on the outside of the cold jar came from the water vapor in the air. How can you be sure that it did not seep out of the jar? The water vapor in the air was cooled when it came in contact with the jar. It condensed and turned into water, leaving the air less humid.

A single jar of ice cubes cannot do much in a spaceship. But cold coils, such as you have in a refrigerator or in an air conditioner, can condense a lot more water vapor. Not only can electric cooling coils do a better job in freeing the air of water vapor, but they can be set to go on and off to maintain just enough vapor in the air to be comfortable.

An air conditioner in your home lets condensed water run off outside. But in a spacecraft, the water can be used again and again. Condensed water is pure water and perfectly good for drinking or for any other needs.

Of course, using condensed water again and again is nothing new. It happens all the time right here on the planet earth. Do you know how all of us depend on water evaporating and condensing? Lakes, streams, rivers, oceans, clouds, and rain are all parts of a large spacecraft. We call that large spacecraft the planet earth.

Replacing Oxygen
in a Spacecraft

YOU WILL NEED: A wide-mouthed jar, water, vinegar, two dry cells or two flashlight batteries, two worn-out flashlight batteries, bell wire, electrical tape, two pyrex test tubes with corks to fit, matches, a thin sliver of wood, a measuring cup, and a ruler.

WHAT TO DO: *Caution: Work this experiment in the kitchen sink because it involves lighted matches.* Break open the worn-out flashlight cells and remove the carbon rods. Wash these rods clean. Cut two 24-inch pieces of wire and bare all four ends. Tightly wrap a piece of the bare wire around each of the brass caps on each rod. Tape the other end of each wire to the taped-together good flashlight cells as shown in the drawing. Half fill the wide-mouthed jar with water and add a cup of vinegar. Insert the carbon rods in the jar and cover each rod with a test tube. Wait for a few minutes until bubbles start to form on each rod. One rod will have twice as many bubbles as the other. These are bubbles of hydrogen. The other rod has bubbles of oxygen. When gas starts to bubble out from the open end of the test tube, carefully lift the tube off the rod, placing your thumb over the open end under water to cap the tube. Remove the tube from the water and insert a cork. Light the end of the wood sliver and blow it out. Quickly remove the cork and insert the glowing tip into each test tube in turn. *Caution: Point the open end of the test tubes away from your body.*

WHAT TO LOOK FOR: Water contains two parts of hydrogen and one of oxygen. Adding the vinegar to the water allows an electric current to flow through it.. The flow of electricity breaks the water down into the two gases, hydrogen and oxygen. This process is called electrolysis.

You use about 3 pounds of oxygen in a day's breathing. For a short trip to the moon, you could carry the oxygen you need aboard the spacecraft in special tanks. But for longer trips into space, we have to find a way to use the oxygen again and again.

One way is to use the water vapor you normally breathe out. The oxygen you inhale goes to every cell in your body. In the cells, the oxygen combines with certain materials, mainly carbon and hydrogen. These combinations form carbon dioxide and water.

The water vapor in your breath can be condensed and then the oxygen present in the water can be collected by electrolysis. In this way, the oxygen is regained and can be used again and again. Can you think of a use for the hydrogen being collected?

Removing Carbon Dioxide
from a Spacecraft

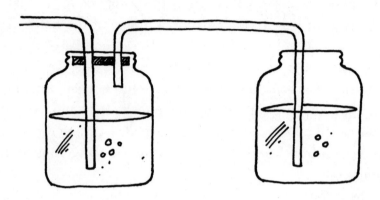

YOU WILL NEED: Calcium hydroxide or calcium oxide (inexpensive chemicals found in many chemistry sets), water, a glass, a drinking straw, 3 or 4 jars with covers, plastic tubing, and a two-holed rubber stopper to fit the jar top (or you can use some child's plastic clay).

WHAT TO DO: Add enough calcium hydroxide or calcium oxide to the water in a jar so that some is left on the bottom after the rest dissolves. Close the jar tightly and allow it to stand for a day. When you return after a day, pour off the clear liquid at the top into another jar and keep this jar tightly closed. The clear solution you have is called limewater. You can discard the contents of

53

the first jar. Pour a little bit of the limewater into a glass. With the drinking straw, bubble your breath into the limewater. After a while, it should turn milky white. Put the glass containing the limewater aside for later observation.

Next set up the other two jars as in the diagram, placing some limewater in each jar. Place the rubber stopper in one jar, and bubble your breath through a tube into one of the holes. A tube from the second hole should lead into the second jar of limewater.

WHAT TO LOOK FOR: When we breathe, we exhale carbon dioxide. Carbon dioxide is not harmful in small amounts. On the earth, plants use carbon dioxide to make food and oxygen in a process called photosynthesis. In a spacecraft, carbon dioxide must be prevented from building up and becoming dangerous to the astronauts.

The carbon dioxide in your breath reacted with the limewater. That is what made the limewater turn milky white. After a while the white material settles to the bottom of the glass. The carbon dioxide you breathed out is now part of that white solid material. The white material is called calcium carbonate. The chemical reaction that took place is this: calcium hydroxide + carbon dioxide = calcium carbonate + water.

Compare the color of the limewater in the two jars. Which shows more of a reaction? How does this show that limewater can dispose of carbon dioxide? How could you dispose of even more carbon dioxide? Why is this method preferable to using green plants on short space flights?

54

A Closed Life-Support System

Y O U   W I L L   N E E D : A clean wide-mouthed jar with a plastic cover, water plants such as elodea, two or three pond snails, water, and candle wax.

W H A T   T O   D O : Make sure that the jar is clean. Rinse it free of all traces of soap. Fill the jar almost to the top with pond water or water from an aquarium. If neither of these sources are available, draw water from a tap and let it stand in an uncovered container for at least 48 hours. Place several sprigs of elodea in the water along with the pond snails. Make sure that the water plant

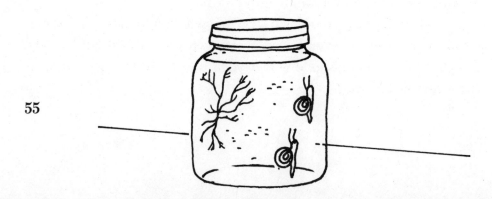

is green and firm. A rotting plant is soft to the touch and a darker green color. It will decay very quickly in the jar. Use a tight-fitting cover to close the jar. Seal all around the edges with melted candle wax. Keep the jar in a well-lighted place near a window, but not in direct sunlight for any length of time.

WHAT TO LOOK FOR: How would it be possible for man to make trips of several months or even years to distant planets? How could enough food and oxygen be carried in a spacecraft of a reasonable size?

One way would be to design a closed system containing plants as well as the astronauts. Briefly, the system would work around the idea of the plants supplying some of the food and oxygen, while the astronauts would supply the carbon dioxide and other materials needed by the plants.

In the closed-system aquarium that you have set up, the plants and the snails each provide something that the other needs. You may see bubbles of gas forming on the leaves of the plants during the day. The gas is oxygen. The snails use the oxygen and eat the plant material. In turn, the snails give off carbon dioxide and wastes that are used by the plants for growth and to make more oxygen.

One kind of plant that is being considered for long space trips is a microscopic green algae called chlorella. Under intense light, the plant multiplies very rapidly. When taken from the water and dried, chlorella can be made into a powder that can be used as a source of food. During growth, chlorella gives off a great deal of oxygen. Still another plant that is being considered for long trips is the common corn plant.

Of course there are many problems to solve before a spacecraft could be made self-sustaining. Can man live on the plants grown on such a trip for a long period of time? Wouldn't such a diet become very monotonous even if it did contain enough nutrients? How many such plants and how much equipment would be necessary to set up such a closed life-support system? The chances are that such a system will not be used until man stays in space for very long periods of time. But since man has solved the problems of going into space, he can probably overcome the difficulties in making a life-support system.

Psychological Aspects
of Space Travel

Y O U W I L L N E E D : You, your friends, small animals such as mice or hamsters, mazes, and a phonograph turntable.

W H A T T O D O : Make a list of the conditions that man must have if he is to live in space. How does he go about solving the problems of air, temperature, food, gravity, radiation, and so on? But what about the other effects that space may have? What about the problems of noise and shaking during launch and reentry? What about the loneliness in a spacecraft or on the moon? What happens when you are crowded into a small space with a group of other people? Do you ever become irritated and angry with them?

58

Could you sleep in definite time slots? Could you stand the sensation of no day or night? Think of how you would go about experimenting with each of these aspects of space travel. Perhaps you can think of a way you could simulate aspects of space travel for you and your friends. Use *Books for Reading and Research* on page 83 for help in finding out some of the training that astronauts must undergo.

WHAT TO LOOK FOR: Try to devise a test that you can take, such as a large group of math questions in addition and multiplication. Take the test under normal conditions and compare the results under the conditions of space travel that you set up. For example, how would you do on this kind of test after staying in isolation in a very small, cramped room? How would you do on this test after spinning around for a while? Think of other ways you could test the effects of your space simulations.

You might use animals in the following way. Train a mouse or hamster to run through a maze. Record the average time he takes after the maze is learned. Place the animal under some condition that you are testing, such as spinning it in a cage for several minutes on a phonograph or keeping it under constant lighting for a day. Then test it in the maze again. Compare the maze performance after experiencing these conditions with the performance under normal conditions. *Caution: Be sure not to do anything to an animal that might injure it.*

After you have tried out a number of these tests, do you think that long, manned space flights are practical? What conditions must an astronaut have to stay in space for a long time? How long could crews remain on a space station or on the moon? Would you or your friends like to go on a space trip for days? For weeks? For years? What kind of people would volunteer for a space trip to a distant star?

Some of the more distant journeys may take generations. Whole families would go. Youngsters would be born in space and learn how to control the craft and what their mission is. In turn, they would teach their children and their children's children. What do you think of such a trip? Would you go on one?

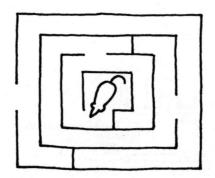

part **3**

How Can We Explore Space?

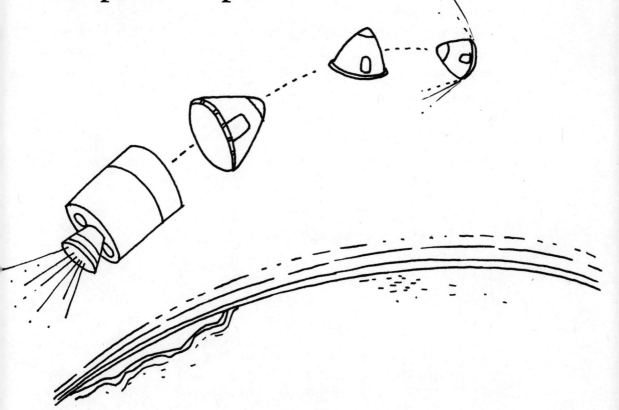

Measuring Distances in Space

YOU WILL NEED: A pencil, a ruler, a large sheet of paper, and tape.

WHAT TO DO: Hold up the pencil at arm's length. Look at the tip with one eye open and the other eye closed. Now look at it with the other eye open and the first eye closed. The pencil will appear to change position against the background.

Draw vertical lines 2 inches apart down the length of the paper. Tape the paper to a wall in your room. Stand about 6 or 8 feet away. Again hold the pencil at arm's length. This time as you open and close each eye, count the number of lines on the paper that the pencil appears to move across. Hold the pencil closer to your eyes and follow the same procedure. Vary the distance and try again.

WHAT TO LOOK FOR: When you travel in a fast-moving car or train, the moon or the sun appears to move along with you. But telephone poles or trees change their positions rapidly. Can you explain why? Do nearby or distant trees appear to move more rapidly or more slowly?

The amount an object appears to move against a stationary background is called parallax. Astronomers can follow and measure the movement of planets and closer stars against the background of more distant stars. Of course, they have to do more than blink their eyes.

Can you tell what would happen to the angle of parallax if you looked from places that were farther apart? Try it and see. From one side of a room, describe what you see as the background to an object in the center of the room. Now move to the other side of the room. What does the background of the object look like now?

Astronomers need to use as large as possible a base line for their parallax observations. Sometimes two observatories on different sides of the earth are used. Then the base line is the distance between the observatories. Other times an even larger base line is used. An observation of a star is made during one month. Six months later when the earth is 186 million miles (the diameter of the earth's orbit) away on the other side of the sun, another observation is made. Even this large base line is too small to observe parallax changes in very distant stars. Other methods involving star types are used. Other than the sun, the star closest to the earth is over 25 trillion miles away. The number of miles to a star is so large that we use another unit of distance. It is called a light-year. Light travels at about 186,000 miles per second. In one year, light will travel a distance of nearly 6 trillion miles. We call that distance, one light-year. Can you figure out how many light-years away the closest star is? Its name is Proxima Centauri.

Sirius, one of the brightest stars in the sky, is 8.6 light-years away. Polaris, the North Star, is 40 light-years away. A cluster of stars in the constellation Hercules is 35,000 light-years away. The Andromeda Galaxy of stars is 2.3 million light-years away. Can you figure out those distances in miles?

Testing the Moon's Surface

YOU WILL NEED: Rice, gravel, different soils, small rocks, a potato, a marshmallow, paper cups, a plastic drinking straw, a small postal scale, and a thin piece of metal tubing.

WHAT TO DO: Try pushing the plastic straw through paper cups each of which is filled with one of the materials above. Push with the flat of your palm and compare how hard each material is. To make a more accurate hardness probe, use the metal tubing. Push against the top of the tubing with the inverted postal scale. Read the amount of push on the scale.

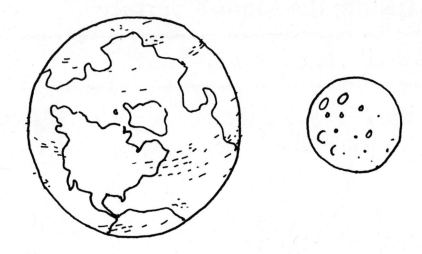

WHAT TO LOOK FOR: The hardness of the moon's surface is one of the things lunar probes and the moon astronauts test for. Scientists need to know about the crust of the moon for many reasons. They would like the answers to questions such as: How and when was the moon formed? Can the moon's surface support the heavy weights of future rockets and lunar bases? What valuable minerals may be found on the moon? In what ways is the moon's surface like that of the earth? In what ways is it different?

What other facts about the moon would scientists like to find out? Make a list of things you would like to know and compare it with the experiments carried out by our lunar probes and astronauts. Use some of the sources listed in *Books for Reading and Research*.

Detecting Life
on Other Planets

YOU WILL NEED: Several baby-food jars with tight-fitting covers, gelatin, water, a large pan, string, an electric hot plate, and a ruler.

WHAT TO DO: Pour about 1 inch of water into the pan. Stand the baby-food jars in the pan and place on the hot plate. Cut the string into 1-foot pieces and place them in the pan as well. Heat the water and let it boil for about fifteen or twenty minutes to sterilize the jars. *Caution: Be careful with boiling water, and protect your hands when picking up the hot jars.* Prepare the gelatin with boiling water as per directions on the package. Pour

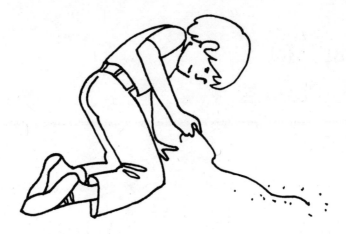

the liquid gelatin into the sterile baby-food jars. Cover them and allow them to cool off. After the strings are cool, take one and drag it across the floor. Place it in a sterilized jar, and cover. Drag another string across the sidewalk or across some soil and place it in another jar. Use a piece of string dragged across a different place in each jar except for one. This jar will be your control. Keep the jars in a warm spot in your home without opening them, and observe them every day for about one week.

WHAT TO LOOK FOR: Here is one way scientists hope to find out if there is life on Mars. It is called Gulliver and it works with sticky strings. When an unmanned space probe lands on Mars, bullets will shoot out sticky strings in different directions. The strings will then be dragged back across the ground and put into a container. Within the container is a special food. If there is microscopic life on Mars, it may multiply quickly within the container. Instruments inside the container will detect the growth and radio the results back to earth.

Of course, there is a question about the experiment. If there is life on Mars will it grow using the same kind of food as life back on earth? Scientists think so. They also think that life will grow at the temperatures found in the probe. But they are not sure.

What about your sticky strings? Do you notice any kinds of life appearing on or around the strings? You may find things such as molds, yeasts, and colonies of bacteria. Do you notice any forms of life growing in the control bottle? What were you trying to make sure of by using a control? Do you think that all the living things that you collected on your strings will grow in the jars?

If life exists on Mars, or on any of the other planets, man wants to find out about it. Space probes are now being designed to detect different kinds of life under different conditions. Some of the designs for life-detectors are even stranger than Gulliver!

Detecting Magnetic Fields in Space

YOU WILL NEED: Two bar magnets, string, a compass, and an empty shoe box.

WHAT TO DO: Tie the string to each end of a bar magnet. Attach the string to some object in the room so that the bar magnet is suspended in the air and can rotate freely. When it stops turning note the direction in which it points. Tap it gently to start it

turning. When it stops, again note the direction. Bring the other bar magnet near the suspended one and see what happens. Finally, put a bar magnet in the shoe box. Move the compass around the outside of the box. What happens to the compass needle? Try to detect the magnetic field around the box.

WHAT TO LOOK FOR: Space probes and manned spacecraft carry instruments to detect magnetic fields around other planets. One such instrument is called a magnetometer. You showed the principle of a magnetometer when you suspended a freely swinging bar magnet. Do you know what other instrument works on the same principle?

The earth is surrounded by a magnetic field. Curved lines can be drawn to make a picture that shows the shape of the field. The lines are called magnetic lines of force. How do these lines of force help to explain why when the bar magnet stops swinging, it always points in the same direction? In what direction does the bar magnet line up? Use the compass to make sure.

71

What happens when you bring the second magnet close to the suspended magnet? Can you tell why? Venus has been found to have a very weak magnetic field, possibly only one-thousandth as strong as the earth's field. Can you tell how a magnetometer in a space probe might have discovered this?

When Mariner IV flew close to Mars, its magnetometer did not change position. Can you tell what that showed? Other planets such as Jupiter may have a very strong magnetic field. Many astronomers believe that there are magnetic fields in space between the stars. How will we find out about these things?

What happens when you move the compass around the shoe box containing the bar magnet? Can you think of how you could use the changes in the compass's direction to draw a map of the magnetic field around the box? Try placing the box atop a sheet of paper and drawing lines in the same direction as the compass lines up. Does a magnetic field also extend above and below as well as on a flat surface? Where is the field strongest? Weakest? Try moving the compass all around and see.

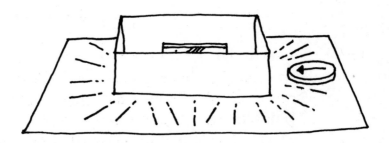

Devising a Space Language

YOU WILL NEED: A pencil and paper, and a friend to help you.

WHAT TO DO: Suppose that there is a civilization on a distant planet that could pick up television signals from the earth. How could you devise a picture language that could be understood by an alien intelligence? As a first step try drawing a series of shapes that could be universally recognized as symbols for

73

numbers. Then think of how you could explain how other shapes stand for addition, subtraction, and so on. Try out what you devise by presenting the series to a friend. Tell him to pretend that it is a message from another civilization and that he has to decode it.

WHAT TO LOOK FOR: Any distant alien civilization that can pick up television signals from the earth should be able to figure out a consistent picture language. Philip Morrison, a Cornell University physicist, has devised such a system. Here, for example, is how he might send the problem $1 + 4 = 5$:

Can you decipher the symbols used? Try writing another problem in the same kind of picture language. Present the problem to a friend and ask him to decode it.

Once the distant television viewer has discovered how to receive the pictures correctly, we could then send out regular pictures that could transmit any information we wanted. One difficulty is: How do we know if someone is listening? How could we recognize a signal that they send out as a reply?

74

Astronomers are training radio telescopes at all parts of the sky. So far they think that they have picked up radio waves from natural sources only. They hope to recognize a radio signal from an intelligent being by its regularity or by some other feature such as a pattern like the one of the space language shown above.

There is still another difficulty in communicating with a planet or a distant star. Radio waves travel at the speed of light, 186,000 miles per second. On earth that is very fast. But even at that speed, radio waves will take years to reach even the closest stars. Can you imagine saying hello — and then waiting ten years for an answer?

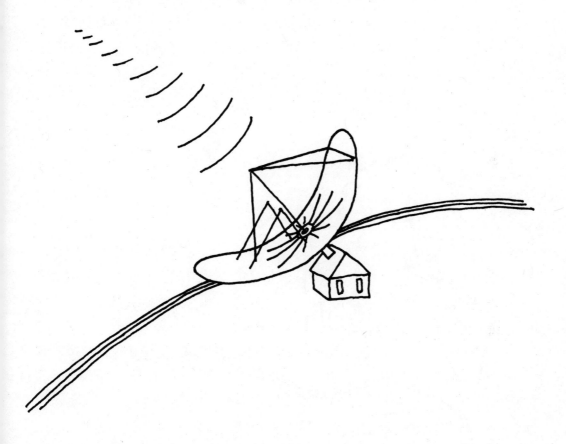

Twinkle, Twinkle, Little Star

YOU WILL NEED: A flashlight, a magnifying lens, an electric hot plate, a piece of waxed paper, and a book.

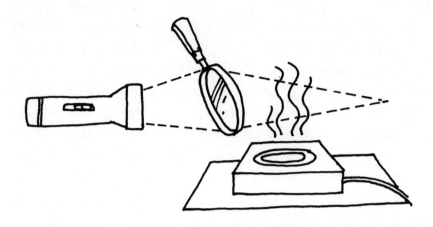

W H A T T O D O : Use the magnifying glass to focus the beam of light from the flashlight onto a wall. Place an electric hot plate or any other heat source just in front of the lens and below it. *Caution: Do not handle the hot plate when it is turned on.* Turn on the heat and observe what happens to the pinpoint image of the light.

For the next demonstration hold the waxed paper close to your eyes and try to read the printing in this book. Now press the waxed paper against the pages of the book and try to read the printing.

W H A T T O L O O K F O R : Rapid, small changes in the brightness or position of a star are called twinkling. The twinkling of stars is caused by the motion of the earth's atmosphere. Quickly moving air currents make observing very difficult. The air currents are caused by differential heating and cooling of the earth's surface. Can you think of why astronomers like to have their observatories built on mountaintops away from populated centers?

77

The view from space is much better than the view from earth. From space, the stars shine steadily against a black background. No atmosphere in space disturbs the passage of light rays. Of course, the same lack of atmosphere is true of the moon. For this reason, a lunar observatory may be built in the not-too-distant future.

Looking back to the earth from the moon or from a satellite, the astronauts are able to make out surface features such as rivers and lakes very clearly. The demonstration you tried with the waxed paper gives you a clue as to why this is so. Think of the waxed paper as the earth's atmosphere. When you changed the position of the same piece of waxed paper from close to your eyes to close to the printed page, what difference did you observe? For you, the position of the waxed paper close to the book is similar to the position of the atmosphere close to the earth for an astronaut out in space. In both cases, what seems to happen when a person looks from outside the distorting layer? Look at some of the photographs of the earth taken from space and check to see if this is true. For sources of photographs check *Books for Reading and Research* on page 83.

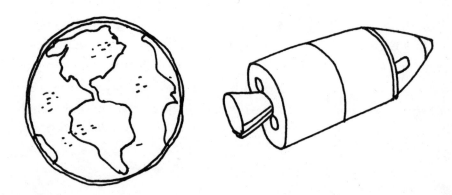

The Color of a Star

YOU WILL NEED: A piece of iron or steel wire, a cork, a ruler, a gas burner on a stove, and a pot holder.

WHAT TO DO: You can purchase iron wire in a hardware store. Picture-hanging wire can also be used. Cut off a 10-inch length of wire. Insert one end into the cork. Make a loop at the other end. Hold the cork end with a pot holder and place the loop into a gas flame. Keep the loop in the flame as it turns colors.

WHAT TO LOOK FOR: This demonstration is best carried on in a darkened room. The heated iron will start to glow. At first it glows red. Then as it gets hotter, the color changes. Which color does it glow at its hottest?

Generally, blue-white stars are the hottest, while red stars are the coolest. Our sun, a yellow star, is somewhere in between these extremes. The spectrum of a star (see page 82) is also related to the temperature of the star. The color and spectrum of a star help astronomers to learn much about the star's age and composition.

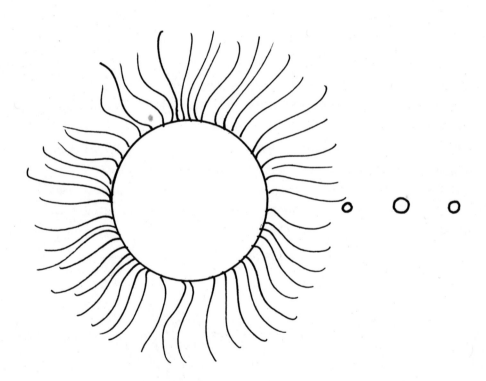

The Composition of a Star

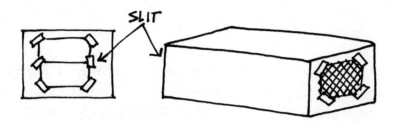

SLIT

YOU WILL NEED: A candle or an alcohol lamp, matches, table salt, Epsom salts, baking powder, a cardboard shoe box, two index cards, tape, a scissors, and a diffraction grating. (You can buy an inexpensive diffraction grating set in cardboard from one of the scientific supplies companies listed in *Where to Buy Supplies*.)

WHAT TO DO: *Caution: Work this experiment in the kitchen sink because it involves lighted matches.* At one end of the box cut a hole a bit smaller than the grating. Tape it over the hole. Cut a ¼-inch-wide hole about 1 inch long at the other end of the box. Tape the two index cards over the hole so that the edges form a very narrow slit. Light the candle and point the slit end of the box at the flame. Look at the flame through the diffraction grating. You should see a color fringe on each side of

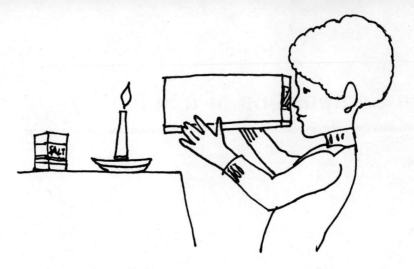

the slit. If the colors appear only at the ends of the slit, untape the grating, turn it 90°, and retape it to the box. Sprinkle each of the substances on the flame as you look through the grating.

WHAT TO LOOK FOR: Each element (when it is heated enough to become a gas) sends out only a certain set of colors. A scientist can recognize an element by the colors emitted when it is burned.

To identify the element, a thin beam of its emitted light is sent through a diffraction grating. The grating separates the colors in the light and forms a spectrum. This kind of spectrum is composed of a series of bright lines spread out according to the colors.

For example, sodium when heated gives off a bright yellow color, calcium gives off an orange-red, and lithium gives off a crimson color. Gas-filled lamps of each of these elements are used to produce different colored lights.

Look at many different sources of light with your diffraction grating. Can you learn to identify the spectrums you see? Do all incandescent bulbs have the same spectrum? Do all fluorescent lights have the same spectrum?

Books for Reading and Research

Barbour, John. *Footprints on the Moon*. New York: Associated Press, 1969.

Bergaust, Erik. *Mars: Planet for Conquest*. New York: Putnam, 1967.

Branley, Franklyn M. *Experiments in Sky Watching*. New York: Crowell, 1967.

Clarke, Arthur C. *Man and Space*. New York: Time, 1968.

Cooper, Henry S., Jr. *Apollo on the Moon*. New York: Dial, 1969.

Ehricke, Kraft and Betty Miller. *Exploring the Planets*. Palo Alto: Silver Burdett, 1969.

Glines, Carroll V. *First Book of the Moon*. New York: Watts, 1967.

Halacy, D. S. *Colonization of the Moon*. New York: Van Nostrand, 1969.

Hirsch, S. Carl. *On Course*. New York: Viking, 1967.

Hyde, Margaret O. *Exploring Earth and Space*. New York: McGraw-Hill, 1967.

Knight, David C. *Meteors and Meteorites: An Introduction to Meteoritics*. New York: Watts, 1969.

Lewis, Claudia. *Poems of Earth and Space*. New York: Dutton, 1967.

Ley, Willy. *Rockets, Missiles, and Men in Space*. New York: Viking, 1968.

Lukashork, Alvin. *Communication Satellites: How They Work*. New York: Putnam, 1967.

Newell, Homer E., Jr. *Space Book for Young People*. New York: McGraw-Hill, 1968.

Silverberg, Robert. *The World of Space*. New York: Meredith, 1969.

Wilfred, John N. *We Reach the Moon*. New York: Bantam, 1969.

Young, Richard S. *Life Beyond the Earth*. Palo Alto: Silver Burdett, 1969.

Where to Buy Supplies

Centuri Engineering Company P.O. Box 1988 Phoenix, Arizona 85001	Model rocketry supplies; and a periodical, *American Rocketeer,* distributed free on request.
Edmund Scientific Company 300 Edscorp Building Barrington, New Jersey 08007	All kinds of scientific supplies for amateur astronomers, etc. Write for catalog.
Estes Industries, Inc. P.O. Box 227 Penrose, Colorado 81240	Model rocketry supplies. Write for catalog.
Flight Systems, Inc. P.O. Box 145 Louisville, Colorado 80027	Model rocketry supplies. Write for catalog.
Hubbard Scientific Company 2855 Shermer Road Northbrook, Illinois 60062	Scientific supplies.

Index

composition of, 82
distance from earth, 64
measurement of movement of, 63
observation of, 76-78
spectrum of, 80
temperature of, 80

Television, role in space communication, 73-74
Thrust, *see* Rocket thrust

Velocity
  effect of force on, 6, 7, 8
  effect of mass on, 7-8

effect of mass ratio on, 15
escape, 19-21
horizontal, 5
measurement in spacecraft of, 40-41
and weightlessness, 35
Venus, magnetic field of, 72

Water
  as propellant, 14-16
  in spacecraft, 49
Water rockets, 14-16
Water vapor, in spacecraft, 48-49, 52
Wax, as heat shield, 28
Weightlessness, 34-36

ABOUT THE AUTHOR

Seymour Simon has taught science on a number of levels for over ten years. He is at present teaching that subject in a New York City school. Mr. Simon is the author of *Animals in Field and Laboratory, Exploring with a Microscope,* and *Science at Work: Easy Models You Can Make.*